T0237441

SpringerBriefs in Pharmaceutical Science & Drug Development

SpringerBriefs in Pharmaceutical Science & Drug Development present concise summaries of cutting-edge research and its applications in the fields of the pharmaceutical and biotechnology industries. Featuring compact volumes of 50 to 125 pages, the series covers a range of content from professional to academic. Typical topics might include:

- A timely report of state-of-the art analytical techniques
- A bridge between new research results, as published in journal articles, and a contextual literature review
- A snapshot of a hot or emerging topic
- An in-depth case study or clinical example
- Practical information and background to allow researchers to transition into new areas

SpringerBriefs in Pharmaceutical Science & Drug Development will allow authors to present their ideas and findings to their scientific and research communities quickly and will allow readers to absorb them with minimal time investment. Both solicited and unsolicited manuscripts are considered for publication in this series.

More information about this series at http://www.springer.com/series/10224

Thomas Catalano

Application of Project Management Principles to the Management of Pharmaceutical R&D Projects

 Springer

Thomas Catalano
N/A
PharmChem Analytical Consultants
Buffalo Grove, IL, USA

ISSN 1864-8118   ISSN 1864-8126   (electronic)
SpringerBriefs in Pharmaceutical Science & Drug Development
ISBN 978-3-030-57526-7   ISBN 978-3-030-57527-4   (eBook)
https://doi.org/10.1007/978-3-030-57527-4

This Springer imprint is published by the registered company Springer Nature Switzerland AG
The registered company address is: Gewerbestrasse 11, 6330 Cham, Switzerland

*This book is dedicated to my wife Jeanie
and my four daughters
Margaret, Linda, Jeanmarie, and Michelle
For their support during my writing endeavors*

Preface

Within the last 30 years Project Management has become a major part of the commercial business processes throughout the world. Project Management utilizes sophisticated computer software to apply the system. In fact, Project Management has become a discipline of its own, where universities are now giving bachelor's and master's degrees in this subject. During my 10 years of consulting, I have come to realize that small businesses, generic companies, start-ups and virtual companies do not have the budget or the trained resources to apply the software. Therefore, they neglect using the principles of Project Management in their business model, which reduces their efficiency and increases their operating costs. I saw this as an opportunity to help these businesses take advantage of the benefits of using the Project Management system in their business model. I decided to author a book titled *Application of Project Management Principles to the Management of Pharmaceutical R&D Projects*.

Within this book the Project Management Concepts is presented as a paper-based system for completing all the critical activities needed for Project Management. Focusing on a paper-based system will allow those businesses who don't have access to the software to still utilize the project management system. This will allow these companies to take advantage of all the principles of Project Management and gain all the advantages associated with the system.

Buffalo Grove, IL, USA   Thomas Catalano

Contents

Abbreviations

AD   Analytical Department
AL   Analytical Lead
AMD   Analytical Method Development
AMV   Analytical Method Validation
CDA   Confidentiality Agreement
CMC   Chemistry Manufacturing and Controls
CofA   Certificate of Analysis
DF   Dosage Form
DP   Drug Product
DS   Drug Substance
GLP   Good Laboratory Practices
GMP   Good Manufacturing Practices
HLA   Highest Level of Authority
ICH   International Council for Harmonisation
ID   Identity, Identification
LOQ   Limit of Quantitation
MSDS   Material Safety Data Sheet
NAC   Nominal Analytical Concentration
OQ   Operational Qualification
PC   Process Chemistry
PH   Phase
PR   Project Research
PQ   Performance Qualification
PSA   Product Safety Assessment
QA   Quality Assurance
QC   Quality Control
$\%R_c$ Percent risk change between plans
$\%R_t$ Percent total risk of current plan
$\%R_{XL}$ Percent risk evaluation of determined low risk plan
Ref. Std. Reference Standard
RFI   Request for Information

ROA	Report of Analysis
SOP	Standard Operating Procedure
Tox.	Toxicology
x_c	Total risk value for current plan
x_L	Total risk value for determined low risk plan
x_{max}	Maximum total risk value
x_{min}	Minimum total risk value
x_n	Total risk value for new plan

List of Figures

List of Tables

Chapter 1
Introduction

Abstract In This chapter the reader is introduced to the concepts, implementation and principles of Project Management directed towards small pharmaceutical businesses, such as generic, start-up, and virtual companies. Project Teams, the backbone of project management, its formation, components, team dynamics and member roles and responsibilities are highlighted. A paper base management system is focused on, so that small businesses can take advantage of the project management system in their business model. All of the subject matter described above will be discussed in detail throughout the book.

Introduction

In this manuscript, an organizational structure is proposed for the organization and management of the project team model. Discussions of the team components, management, and the utilization of team dynamics are presented in detail. The organization structure is the starting point for initiating an effective and efficient project management process. The structure should contain process owners or expert groups to ensure the expertise is available for all stakeholders of the project. The backbone of the project is the **Project Teams**. The teams are resourced from the staff population and exchanges can be made as the project moves through the various stages of development. Within the staff, individuals are identified as team leaders, process experts, team representatives and facilitators. The team leader and representatives make up the project team. Process experts are utilized by the teams (see Fig. 1.1), and can also be part of the team as a team leader or members. The project facilitator is not a team member, their function is to control all aspects of the team dynamics. Individual team leaders, process experts and representatives can contribute to several project teams simultaneously. Quality flow diagrams are presented to demonstrate the project activity flow. A process for project risk evaluation based on the project plan is proposed. The Project Team Model is shown in Fig. 1.2. Application of Project Management Principles to the Management of Pharmaceutical R&D projects is presented as a paper- based system for completing all the critical activities needed for Project Management. The reason for not utilizing all the computer

T. Catalano, *Application of Project Management Principles to the Management of Pharmaceutical R&D Projects*, SpringerBriefs in Pharmaceutical Science & Drug Development, https://doi.org/10.1007/978-3-030-57527-4_1

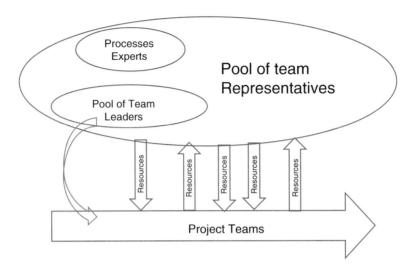

Fig. 1.1 Project team Model

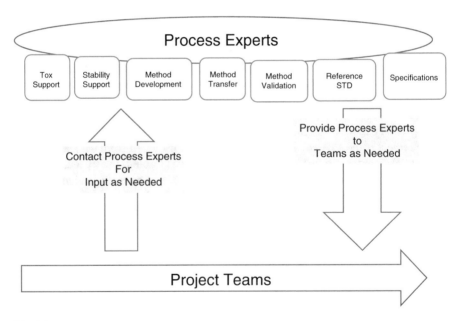

Fig. 1.2 Process experts

software available and focusing on a paper- based system is that many small companies, generic companies, start-ups and virtual companies do not have the budget or the trained resources to apply the software. Therefore, they neglect using the principles of Project Management in their business model, which reduces their efficiency and increases their operating costs. This paper- based approach will allow these companies to take advantage of all the principles of Project Management and gain all the advantages associated with the system.

Chapter 2
Total Quality Management (TQM)

Abstract Project management utilizes the concepts of **Total Quality Management** to institute effective communication and an integrated system to implement a successful project strategy and plan. In this chapter quality management components such as, developing a problem statement, performing brainstorming, asking clarifying questions, making decisions, reaching consensus, and implementing best practices, etc. are discussed and shown how its utilized within a Pharmaceutical R&D project team.

Total Quality Management (TQM) [1]

TQM can be summarized as a management system for a customer-focused organization that involves all employees in continual improvement. It uses strategy, data, and effective communications to integrate the quality discipline into the culture and activities of the organization. These concepts are considered so essential to TQM that they are utilized as a set of core values and principles on which the organization is to operate. Here are the 8 principles of total quality management:

1. Customer-focused: The customer dictates the level of quality required. No matter what is done to foster quality improvement such as, training employees, integrating quality improvement into the process, or upgrading equipment, the customer determines whether the results of the improvements are acceptable.
2. Total employee involvement: All employees are working toward common goals. When empowerment is provided by management the end result is high-performance which enables continuous improvement efforts within the business operations. Project teams are a form of empowerment.
3. Process-centered: A major part of TQM is a focus on process thinking. A process is a series of actions that take inputs from suppliers and transforms them into outputs that are received by customers. The action required to carry out the process are defined, and process performance is continuously monitored in order to detect unwanted variation.

T. Catalano, *Application of Project Management Principles to the Management of Pharmaceutical R&D Projects*, SpringerBriefs in Pharmaceutical Science & Drug Development, https://doi.org/10.1007/978-3-030-57527-4_2

4. Integrated system: Although an organization may consist of many different functional groups that are organized vertically. It is the horizontal interconnecting of these functions that are the focus of TQM. All processes must aggregate into the project processes required for the implementation of the project strategy and plan. Everyone must understand the quality policies, objectives, and critical processes of the organization. The project performance must be monitored and communicated continuously.

5. Systematic approach: A critical part of the management of quality is the systematic approach to achieving an organization's goals. This process, called Project Management, includes the formulation of a project plan that integrates quality as a core component.

6. Continual improvement: A major component of TQM is continuous process improvement. Continuous improvement enables an organization to become more effective at meeting customer requirements.

7. Fact-based decision making: In order to determine an organizations' performance, metric data on performance should be collected. TQM requires that an organization continually collect and analyze data in order to improve their predictions and future actions.

8. Communications: During times of project plan changes, and as part of day-to-day operation. Effective communications are necessary for informing and motivating employees at all levels. Communications usually involve plans, strategies, and timeliness.

Project Management Components

Project Management relies on the Management Concepts of **"Total Quality Management Systems"** and **"The Theory of Constraints"**. The components utilized in Project Management are as follows:

- Construct a problem statement
- Brainstorm causes
- Brainstorm solutions
- Identify constraints
- Rounds of reasoning
- Clarifying questions
- Determine workable solutions
- Choose priority for working on solutions
- Modes of decision making
- Reaching Consensus
- Scribe
- Facilitator
- Implement best practices

Problem Statement

Problem statements are constructed by making proper inquires at the project beginning, examples:

- Are training needs required?
- What is the state of information flow between stakeholders?
- Is essential project information distributed in a timely manner, allowing for the implementation of activities in a timely manner?
- Is there input from the proper stakeholders on setting cost and estimation of timeline for project activities?
- Are there Documented Project Plans available to identify all essential activities, timing, resource commitment and cost etc.?
- Are all contractual agreements reviewed and finalized before the project implementation?
- Has a "Risk Assessment Process" been introduced to evaluate the probability of success due to the initial plan or changes made to the plan

Construction of the Problem Statement [1]

The problem statement is constructed by utilizing input from the inquiries made. These inputs are formatted in a **Fish Bone** diagram shown below in Fig. 2.1. The problem statement should be simple and direct. An example of a problem statement construction for an analytical group function is shown below.

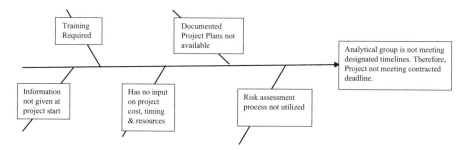

Fig. 2.1 Fish Bone diagram

Project Team Dynamics [2]

There are several team dynamics which should be followed for a team to become a fully functional team. They are:

- Roles and responsibilities of the team leader and members
- Utilization of a brainstorming process
- Rounds of Reasoning
- Clarifying Questions
- Highest Level of Authority (HLA)
- Modes of decision making
- Use of Consensus
- Scribe
- Facilitation
- Governance

The Teams Leader Role Is as Follows

- Act as the leader and a communication link with management
- Set the agenda for each meeting
- Manages the time resource for each agenda Item
- Responsible for the minutes for each meeting
- Is also a member of team as a technical expert in a discipline(s)
- Is the facilitator for the project review meetings

The Team Member Role Is as Follows

- Is the team technical expert in a discipline(s)
- Is also an active contributor to the team outside of their expertise
- May be required to lead sub-teams
- Will make presentations at the project review meetings
- Will contribute to consensus or voting decisions
- Is an effective communicator (written and oral)

Brainstorming Process

The Brainstorming process allows each team member to contribute by making suggestions on activities, issues, solutions etc. in an orderly fashion. The process involves going around the table allow each team member to give their suggestions.

There are no counter arguments allowed from other team members, only clarifying questions could be asked. The Scribe will capture all of the suggestions on flip charts. After several rounds and there appears to be no further suggestions the team leader will end the brainstorming session and move on to a Rounds of Reasoning.

Rounds of Reasoning

Round of Reasoning is a process where each team member is allowed to challenge or support any of the suggestion captured during the brainstorming session. After several rounds, a final list of suggestions is comprised and each team member is given 5 votes which they can place next to the suggestions they support. After all the votes are placed the top 5 selected suggestions are chosen and are worked by the team to obtain a final result.

Clarifying Question

These are questions which are directed towards asking for a better understanding of the issue. Clarifying questions should not be used to pass judgment or disagreement with the issue.

Highest Level of Authority (HLA)

The HLA is generally part of the management team such as; Director, Associate Director, Section Head, etc.

Modes of Decision Making

There are 4 modes of decision:

1. A decision comes from the HLA and the team implements the decision, there is no discussion or feedback from the team.
2. A decision comes from the HLA and feedback is requested from the team. The HLA makes the decision without addressing the feedback given by the team
3. A decision comes from the HLA, there is discussion between the HLA and the team however, the HLA makes the final decision
4. The HLA gives the team complete empowerment to make the decision and the HLA accepts the decision made by the team.

Use of Consensus

The consensus process requires 100% support among the team members; it is not a majority rule scenario like taking a vote. Consensus can still be reached even though members of the team may not fully agree with the decision, but are willing to support the team decision. Consensus is generally better than a voting process because you have the agreement that all team members are willing to support the team decision.

Scribe

The scribe role is to capture important information on to Flip charts, so the information can be placed into minutes or reports. The flip charts are generally kept as the original reference until the activity is finalized. The scribe role is usually shared among the team member based on a schedule. The scribe is still part of the team and should contribute as a team member while acting as the scribe.

Facilitation

The facilitator role is as an overseer of the team process and to help the team to stay on track by making them adhere to the process. The facilitator is not a team member and does not contribute to the agenda. The facilitator has expertise on all team dynamics and total quality management processes.

Governance [3]

Governance is the review of the project by the function management. The department management and the project team make a presentation at the function project team meeting. The function management responds to the presentation with questions and concerns. The department management captures the concerns and any unanswered questions and instructs the project team to evaluate them against the current team project plans. If revisions to the project plans are required, the project team will make the necessary revisions and obtain approval from the department management. The project team will then schedule another meeting with the function management. If no revisions to the project plans are required the department management will prepare a cover letter summarizing the meeting conclusions and attaches it to the finalized presentation which is submitted to the function management.

References

1. Coates and Freeman Inc. (1998) Total quality management consulting. G.D. Searle Inc., Chicago
2. Joiner B 4th Generation management. Rosemount Horizon, Rosemont
3. Pharmaceutical Manufacturers Association (1998) Pharmaceutical management development seminar. Arden House, Harriman

Chapter 3
Theory of Constraints (TOC)

Abstract **Theory of Constraints (TOC)** is another important management tool utilized in the application of project management. TOC is concerned with the flow of activities through the project plan and/or the organization. In every plan there are certain activities that can slow down or stop the flow of the project, these are identified as constraints. In this chapter the focus is on how to identify the constraint and ensure that the activities at the constraint are being performed and that there is no impact on the projects critical path.

Theory of Constraints (TOC) [1]

Theory of constraints is a management concept which is used to manage the flow through a process of interdependent functions for the delivery of a product. There is always at least one constraint, and TOC uses a focusing process to identify the constraint and restructure the rest of the organization around it. TOC adopts the common saying "a chain is no stronger than its weakest link". This means that processes, organizations, etc., are vulnerable because the weakest function can adversely affect the outcome. A constraint in a process can limit the process performance, managing the constraint will result in continuous improvement for the process. There are 4 steps in managing the Constraints

1. **Identifying the constraint** [2] – By asking detailed questions about the process, management and employees working the process should be able to identify where the constraint is located. Depending on the complexity of the process it is easy to miss the constraint location, it's analogous to missing the forest from the trees. Therefore, it is important to ask the right questions. Also, keep in mind that there is always a possibility that there is more than one constraint.
2. **Decide how to optimize the use of the Constraint** – Once the constraint is identified, your first thought should **not** be to eliminate it. You should verify if the constraint is being utilized to full capacity and actions such as idleness at the function and multitasking by process resources is not at the expense of the function where the constraint has been identified.

T. Catalano, *Application of Project Management Principles to the Management of Pharmaceutical R&D Projects*, SpringerBriefs in Pharmaceutical Science & Drug Development, https://doi.org/10.1007/978-3-030-57527-4_3

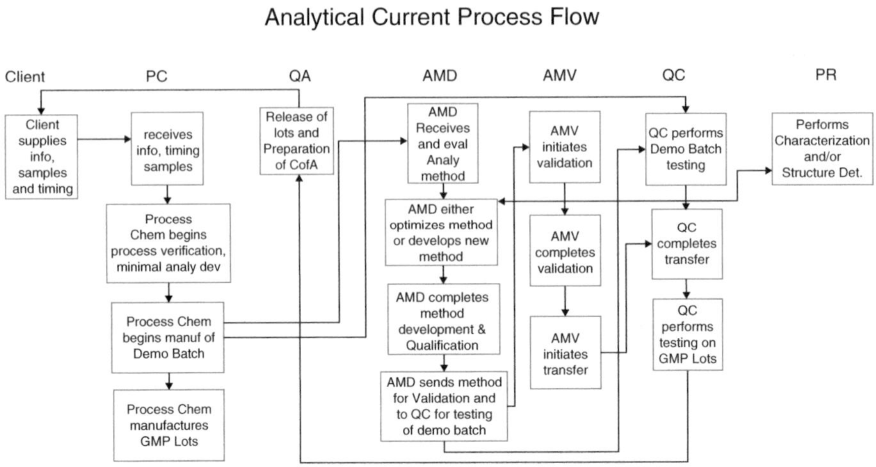

Analytical Current Process Flow

Fig. 3.1 Current process flow diagram

3. **Ensure that the constraint is used optimally** – identify the processes critical path and make sure no time is wasted on that path. Ensure that the process deliverable is acceptable to the client.
4. **Elevate the constraint** – To Elevate the constraint means to invest into increasing resources at the constraint function or the changing of the project plan to eliminate activities, thus keeping the constraint function operating at full capacity. The elimination of activities must be carefully considered since it can increase the project risk. Therefore, a Risk Analysis should be performed before implementing a project plan change.
5. **An example in the use of theory of constraints is shown below:**

> The following "Analytical Process Flow Diagram" represents the **current interaction flow** between Analytical Method Development (AMD) function and other stakeholders within a project (Fig. 3.1).

Observed Constraints

- No interaction between Process Chemistry and AMD at project start
- AMD receives Analytical information, timing and samples at the start of Demo Batch Manufacturing
- Process Chemistry method given to QC at the start of Demo Batch Manufacturing
- Potential backlog of methods to be validated and/or samples to be tested

Identification and resolution of constraint concerns

Analytical Improved Process Flow

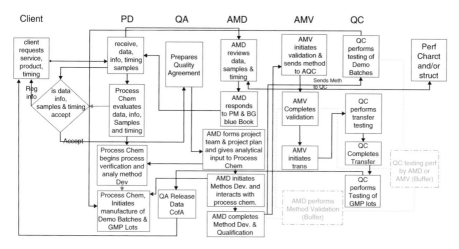

Fig. 3.2 Process flow diagram

- Two constraints were identified

  1. The transfer of information from process chemistry (PC) to AMD in a timely manner
  2. Potential backlog of methods to be validated and/or samples to be tested

- The resolution of constraint 1 was easily resolved by and organizational change, having AMD and PC receiving information, timing and samples at the same time
- Constraint 2 required more evaluation and consideration. The solution is to have all other functions work towards optimizing the capacity at the constrained function. Buffers are also utilized to optimize the capacity at the constraint.

Buffers are the extension of other functions to ensure maximum capacity at the constraint. Observe the use of buffers in the "Analytical Improved Process Flow" diagram Fig. 3.2.

References

1. Uwe Techt (2015) Goldratt and the theory of constraints. ibidem-Verlag Stuttgart, Stuttgart
2. Goldratt EM, Cox J (June 1, 2014) The goal: a process of ongoing improvement paperback. North River Press; 30th Anniversary Edition

Chapter 4
Project Management Process

Abstract This chapter deals with implementation of Project Management Principles (See below) in the management of project teams.

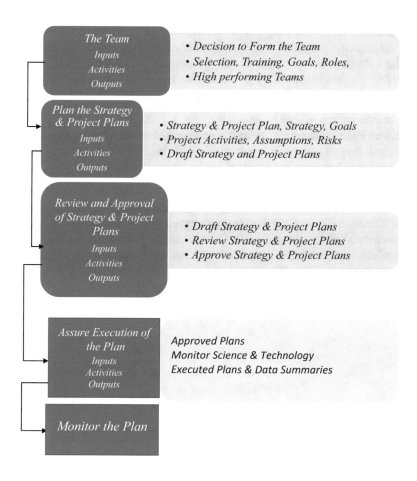

T. Catalano, *Application of Project Management Principles to the Management of Pharmaceutical R&D Projects*, SpringerBriefs in Pharmaceutical Science & Drug Development, https://doi.org/10.1007/978-3-030-57527-4_4

Project Strategy

The development of Project Plan begins with the development of a project strategy [1]. A project strategy checklist is produced (Fig. 4.1) to ensure that all significant activities are included in the strategy plan. A project strategy plan flow diagram (see Fig. 4.2) is developed. The approval of the strategy is shown in Fig. 4.3.

Project Plan

Once the strategy plan is approved, the development of the Project Plan process flow is initiated [1, 2]. The Project Plan directs the project process flow and is critical for a successful outcome. The development of the Project Plan process flow is based on a Project Plan Checklist document (see Fig. 4.4). The Project Plan development process flow is shown in Fig. 4.5 and the Project Plan approval process flow is shown in Fig. 4.6.

- An important activity requirement for the success of a project team, is the **"Project Plan Document"**. Below is an example of the advantage of using the analytical project plan approach to evaluate the initial plan, changes to the initial plan and the associate risks involved by the action taken [3, 4].

Included below is an example of a Project Plan Document template, a populated Project Plan with start dates, finish dates, resources required and precursor links. Also included is a Revised populated project Plan based on the timing and resources demand changes. To evaluate the effect of the change demands in the project Plan, a risk analysis was performed to compare risk between original project plan and the revised project plan. Thus, allowing all stakeholder to understand the details of the project and the effect of the changes requested.

I. Dosage Form

Dosage Form Type	Development Timing	Resource Commitment

II. Pharmaceutics/Pre-formulation

Activity	Development Timing	Resource Commitment

III. Safety Assessment Plan

Tox. Study	Study Timing	Resource Commitment

IV. Analytical Support

Activity	Development/Testing Timing	Resource Commitment

Fig. 4.1 Project strategy checklist

V. Drug Substance

Needs	Timing	Resource Commitment

VI. Outsourcing

Activity	Timing	Resource Commitment	Est. Cost

VII. Documentation

Document	Timing	Responsibility

VIII. Technology Transfer

Activity	Timing	Responsibility

IX. Stability

Study	Timing/Duration	Responsible Site	Outsource Y/N

X. Regulatory Meeting

Meetings	Timing Resource Commitment

Fig. 4.1 (continued)

Project Strategy Development Process

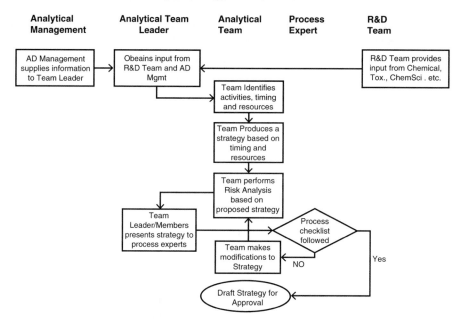

Fig. 4.2 Project strategy

Project Strategy Approval Process

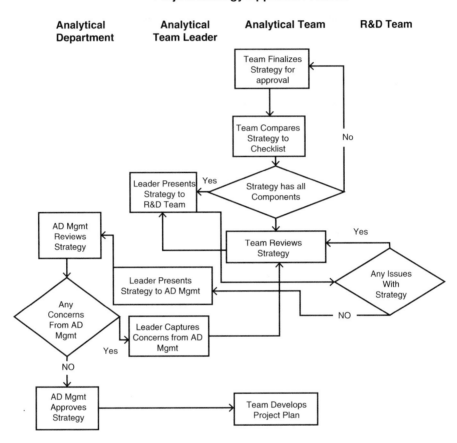

Fig. 4.3 Strategy approval

I. R&D Plan   **Comments**

a. Toxicology Plan

b. Regulatory Plan

c. Documentation Plan (CMC)

d. Marketing Plan

II. Chemical Plan

a. Chemical Availability

b. Production Schedule

c. ID. Comm. Manufacturing Process (Date)

d. Impurities Identification and Qualification

e. Solid State Characterization

f. Stability Plan

g. Release Testing

h. Specification Setting

III. Dosage Form Plan
a. Development of Dosage Form
    Phase I
    Phase II
    Phase III/commercial
    Additional DF development
b. Specification Setting

c. Release Testing

IV. Analytical Technology Plan

a. Development and validation of
  Analytical Methods
b. Prepare Technical Reports

c. Support Specifications Setting

d. Manage Stability Program

e. Support Formulation and Process
Chemistry
f. Release Testing

Fig. 4.4 Project plan development checklist

V. Registration Plan

a. Preparation of Documents (CMC)

b. Finalize Reports

c. Identify Submission Date

d. Respond to Regulatory Responses

VI Technology Transfer

a. Chemical Transfer

b. Dosage form Transfer

c. Analytical Methods Transfer

d. Create the Technology Transfer Team

e. Materials Sourcing Plan

VII Finished Drug Product

a. Release Methods/ Testing

b. Packaging Plan

c. Shipping Plan

Fig. 4.4   (continued)

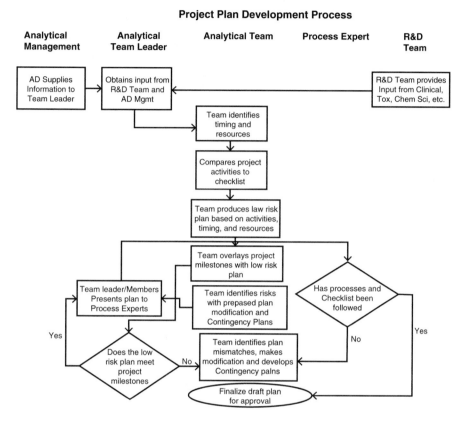

Fig. 4.5 Plan development

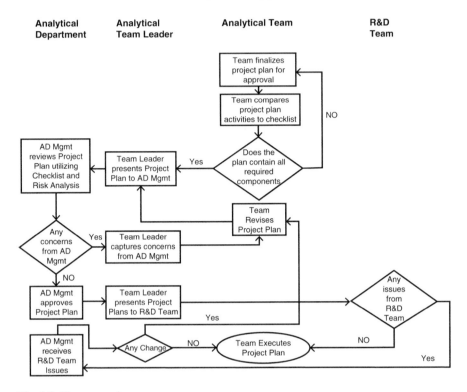

Fig. 4.6 Plan approval

Project Plan Document (Template)

Item number	Activity	Start date	End date	Actual completion date	Applied resources	Precursor links
1	Create a project plan					
2	Pre-method development activities					
	a. Literature search					
	b. Obtain chemical and physiochemical data					
	c. Collect available samples for method development					
	d. Determine method criteria based on stage of development					
	e. Determine chromatography mode					2a,2b,2d

(continued)

Item number	Activity	Start date	End date	Actual completion date	Applied resources	Precursor links
	f. Evaluate mobile phase properties					2b
3	Method development				·	2
	a. Mobile phase screening and buffer using gradient					
	b. Column screening					
	c. Determine whether isocratic or gradient					3a, 3b
	d. Setup solvent selectivity triangle					3a, 3b
	e. Run each of the 10 experiments in the selectivity triangle					
	f. Determine portion of triangle which gives greatest number of peaks and best selectivity					3e
	g. Optimize mobile phase					3f
4.	Sample preparation					
	a. Determine sample diluent					
	b. Determine sample solubility					
	c. Determine extraction efficiency if required					
	d. Determine analyte stability in diluent					
	e. Determine nominal analytical concentration (NAC) and injection volume					
5	Method qualification					
	a. Linear range					
	b. Recovery					
	c. Sensitivity					
	d. Precision					
	e. Initial system suitability parameters					
	f. Development batch analysis					
6	Method validation					
	a. Write a draft method document					
	b. Generate a method validation protocol					5a-e
	c. Approve the method validation protocol					6a
	d. Implement validation experiments					6c
7	Method transfer					

(continued)

Item number	Activity	Start date	End date	Actual completion date	Applied resources	Precursor links
	a. Determine type of transfer					
	b. Prepare intent to transfer document					
	c. Write transfer protocol					
	d. Implement transfer analysis					
8	GMP testing					
9	Technical reports					
	a. Write an analytical development report					3,4
	b. Write a method validation report					6
	c. Reports of analysis (C of A)					8
	d. Method transfer report					7

Example Current Analytical Project Plan (Populated)

Item number	Activity	Start date	End date	Actual completion date	Applied resources	Precursor links
1	Implement a project plan	1/2/18	2/24/18			
2	Pre-method development activities	1/2/18	1/15/18			
	a. Literature search	1/3/18	1/6/18		0.5C	
	b. Obtain chemical and physiochemical data	1/3/18	1/10/18		A	
	c. Collect available samples for method development	1/3/18	1/8/18		B	
	d. Determine method criteria based on stage of development	1/10/18	1/12/18		A	
	e. Determine chromatography mode	1/12/18	1/14/18		A	2a,2b,2d
	f. Evaluate mobile phase consideration	1/14/18	1/15/18		A	2b
3	Method development	1/15/18	2/4/18			2
	a. Mobile phase screening and buffer using gradient	1/15/18	1/17/18		A	
	b. Column screening	1/18/18	1/21/18		A	
	c. Determine whether isocratic or gradient	1/21/18	1/25/18		A	3a, 3b

(continued)

Item number	Activity	Start date	End date	Actual completion date	Applied resources	Precursor links
	d. Setup solvent selectivity triangle	1/25/18	1/27/18		A	3a, 3b
	e. Run each of the 10 experiments in the selectivity triangle	1/27/18	1/31/18		A	
	f. Determine portion of triangle which gives greatest number of peaks and best selectivity	2/1/18	2/1/18		A	3e
	g. Optimize mobile phase	2/2/18	2/4/18		A	3f
4.	Sample preparation	1/13/18	2/5/18			
	a. Determine sample diluent	1/13/18	1/13/18		B	
	b. Determine sample solubility	1/14/18	1/14/18		B	
	c. Determine extraction efficiency if required	NR	NR			
	d. Determine analyte stability in diluent	1/15/18	1/18/18		0.5B	
	e. Determine nominal analytical concentration (NAC) and injection volume	2/4/18	2/5/18		A	
5	Method qualification	2/6/18	2/11/18			
	a. Linear range	2/6/18	2/10/18		A	
	b. Recovery	2/6/18	2/10/18		A	
	c. Sensitivity	2/6/18	2/10/18		A	
	d. Precision	2/6/18	2/10/18		A	
	e. Initial system suitability parameters	2/11/18	2/11/18		A	
	f. Development batch analysis	2/11/18	2/14/18		C	
6	Method validation	2/6/18	2/18/18			
	a. Write a draft method document	2/6/18	2/7/18		B	
	b. Generate a method validation protocol	2/7/18	2/9/18		B	5a-e
	c. Approve the method validation protocol	2/9/18	2/11/18		0.5A, 0.5B	6a
	d. Implement validation experiments	2/11/18	2/18/18		A	6c
7	Method transfer	2/11/18	2/18/18			
	a. Determine type of transfer	2/11/18	2/11/18		B	
	b. Prepare intent to transfer document	2/11/18	2/12/18		B	

(continued)

Item number	Activity	Start date	End date	Actual completion date	Applied resources	Precursor links
	c. Write transfer protocol	2/12/18	2/15/18		B	
	d. Implement transfer analysis	2/15/18	2/18/18		B	
8	GMP testing	2/18/18	2/23/18		0.5B, C	
9	Technical reports	Phase 2B	NDA			
	a. Write an analytical development report	2/21/18	2/24/18		A	3,4
	b. Write a method validation report	2/18/18	2/21/18		A	6
	c. Reports of analysis (C of A)	2/23/18	2/24/18		C	8
	d. Method transfer report	2/18/18	2/19/18		B	7

- Each letter A, B, C, D, etc. in the document, identifies an individual resource. The indication of a partial use of a resource is indicated as a fraction of a whole resource (ex. 0.5A). Precursor Links indicate activities which have direct effect on the activity where the precursor link in located. If changes to the project such as timing, cost, resource commitment arise the current plan may need revision. The use of the project plan document is an ideal document to determine where appropriate changes could be made.

Example Revised Analytical Project Plan (Populated)

Item number	Activity	Start date	End date	Actual completion date	Applied resources	Precursor links
1	Implement a project plan	1/2/18	2/10/18			
2	Pre-method development activities	1/2/18	1/8/18			
	a Literature search	1/3/18	1/6/18		0.5C	
	b Obtain chemical and Physiochemical data	1/3/18	1/10/18		A	
	c. Collect available samples for method development	1/3/18	1/8/18		B	
	d. Determine method criteria based on stage of development	1/3/18	1/5/18		A	
	e Determine chromatography mode	1/12/18	1/14/18		A	2a,2b,2d

(continued)

Item number	Activity	Start date	End date	Actual completion date	Applied resources	Precursor links
	f. Evaluate mobile phase consideration	1/5/18	1/6/18		A	~~2b~~
3	Method development	1/6/18	1/18/18			2
	a. Mobile phase screening and buffer using gradient	1/6/18	1/7/18		A	
	~~b. Column screening~~	~~1/18/18~~	~~1/21/18~~		~~A~~	
	c. Determine whether isocratic or gradient	1/7/18	1/7/18		A	3a, 3b
	d. Setup solvent selectivity triangle	1/7/18	1/9/18		A	3a, 3b
	e. Run each of the 10 experiments in the selectivity triangle	1/10/18	1/14/18		A	
	f. Determine portion of triangle which gives greatest number of peaks and best selectivity	1/15/18	1/15/18		A	3e
	g. Optimize mobile phase	1/16/18	1/18/18		A	3f
4.	Sample preparation	1/13/18	1/17/18			
	~~a. Determine sample diluent~~	~~1/13/18~~	~~1/13/18~~		~~B~~	
	b. Determine sample solubility	1/13/18	1/13/18		B	
	c. Determine extraction efficiency if required	NR	NR			
	d. Determine analyte stability in diluent	1/14/18	1/17/18		0.5B	
	~~e. Determine nominal analytical concentration (NAC) and injection volume~~	~~2/4/18~~	~~2/5/18~~		~~A~~	
5	Method qualification	~~2/6/18~~	~~2/11/18~~			
	~~a. Linear range~~	~~2/6/18~~	~~2/10/18~~		~~A~~	
	~~b. Recovery~~	~~2/6/18~~	~~2/10/18~~		~~A~~	
	~~c. Sensitivity~~	~~2/6/18~~	~~2/10/18~~		~~A~~	
	~~d. Precision~~	~~2/6/18~~	~~2/10/18~~		~~A~~	
	~~e. Initial system suitability parameters~~	~~2/11/18~~	~~2/11/18~~		~~A~~	
	f. Development batch analysis	1/17/18	1/20/18		C	
6	Method validation	1/18/18	1/29/18			
	a. Write a draft method document	1/14/18	1/17/18		0.5B	
	b. Generate a method validation protocol	1/17/18	1/19/18		B	5a-e

(continued)

Item number	Activity	Start date	End date	Actual completion date	Applied resources	Precursor links
	c. Approve the method validation protocol	1/19/18	1/21/18		0.5A, 0.5B	6a
	d. Implement validation experiments	1/21/18	1/29/18		A	6c
7	Method transfer	1/30/18	2/02/18			
	a. Determine type of transfer	2/11/18	2/11/18		B	
	b. Prepare intent to transfer document	2/11/18	2/12/18		B	
	c. Write transfer protocol	1/21/18	1/24/18		B	
	d. Implement transfer analysis	1/30/18	2/02/18		B	
8	GMP testing	2/03/18	2/08/18		0.5B, C	
9	Technical reports	Phase 2B	NDA			
	a. Write an analytical development report	1/30/18	2/02/18		A	3,4
	b. Write a method validation report	2/03/18	2/06/18		A	6
	c. Reports of analysis (C of A)	2/09/18	2/10/18		C	8
	d. Method transfer report	2/03/18	2/04/18		B	7

References

1. Catalano T (2013) Essential elements for a GMP Analytical Chemistry Department. Springer, New York
2. Kerzner H (2001) Project management: a systems approach to planning, scheduling, and controlling, 7th edn. Wiley, New York
3. Catalano T (2001) Pharmaceutical and Analytical Science Team (PAST). G.D. Searle Training Document, New York
4. Coates and Freeman Inc. (1998) Total Quality Management Consulting. G.D. Searle Inc., Chicago

Chapter 5
Risk Evaluation Process

Abstract During the Pharmaceutical Product Development process there is usually a limited amount data available and a limited amount of time to make decisions that could have significant impact on the development of a new chemical entity. Therefore, a Risk Evaluation Process was developed to evaluate risk based on changes to timing, resources, budget, etc. Thereby, allowing the accurate evaluation of the impact of the changes being considered.

Risk Evaluation Process

There are many approaches to risk evaluation such as, Monte Carlo Simulation, Binomial Process, Poisson Process, and Hypergeometric Process to name a few.

All of these processes require sophisticated computer software and a large data set to deliver accurate Risk Predictions [1].

During the Pharmaceutical Product Development process there is usually a limited amount data available and a limited amount of time to make decisions that could have significant impact on the Development of a new chemical entity.

This in-house Comparative Risk Evaluation process was developed not to give absolute Risk Predictions, but more importantly to give an accurate comparative risk evaluation based on changes made in the project plans, due to issues arising during Product Development, such as timing changes, resource changes, budget allocations, etc.

The comparative Risk Evaluation Analysis utilizes weighted activities for each stage of development and a rate factor, which is the probability for the successful completion of the activity in the given time frame.

The product of the weighted activity and the rate factor results in a Total Risk Value. The absolute value of The Total Risk is of minimal significance, it is the new plan Risk Value ($\%R_n$) that is utilized to evaluate the acceptability of the new plan over the current plan [4] (Table 5.1).

T. Catalano, *Application of Project Management Principles to the Management
of Pharmaceutical R&D Projects*, SpringerBriefs in Pharmaceutical Science &
Drug Development, https://doi.org/10.1007/978-3-030-57527-4_5

Table 5.1 Calculate the risk evaluation using the following equations

1.	Total risk evaluation of current plan	$$(\%\mathbf{R_t}) = \frac{(x_c - x_{min})}{x_{max} - x_{min}} \times 100$$
2.	Total risk evaluation of new plan	$$(\%\mathbf{R_t}) = \frac{(x_n - x_{min})}{x_{max} - x_{min}} \times 100 \,(\mathbf{New\,Plan})$$
3.	Risk change index	$$(\%\mathbf{R_{Ch.}}) = \frac{(x_n - x_c)}{x_{max} - x_{min}} \times 100$$
4.	Low risk plan evaluation	$$(\%\mathbf{R_{XL}}) = \frac{(x_L - x_{min})}{x_{max} - x_{min}} \times 100$$
5.	%probability of success (%R$_s$)	$$(\%\mathbf{R_t}) = \frac{x_{c/n} - x_{min}}{x_{max} - x_{min}} \times 100$$
		$$(\%\mathbf{R_s}) = 100\,\% - (\%\mathbf{R_t})$$

Definitions:

x_{max} – Maximum total risk value (all weight activities x5)

x_{min} – Minimum total risk value (all weights activities x1)

x_L – Total risk value for determined low risk plan

x_c – Total risk value for current plan

x_n – Total risk value for new plan

%R$_t$ – Percent total risk of current/new plan

%R$_{chi}$ – Percent risk change between plans

%R$_{XL}$ – Percent risk evaluation of determined low risk plan

%R$_s$ – Percent Probability of Success

$x_{max} - x_{min}$ – The difference between the theoretical max. value and the theoretical min. value

Risk Evaluation Process

Analytical Development

Risk factors	Weightings (1–10)		
	Early stage	Mid stage	Late stage
Method development	3	10	3
Method validation	2	10	10
Availability of lots (DS &DP)	4	8	8
PSA support	7	9	9
Release testing	3	8	10
QC release	3	8	10

(continued)

Risk factors	Weightings (1–10)		
	Early stage	Mid stage	Late stage
Manufacturing sites involvement	2	7	10
Specifications	1	3	10
Methods for alternative DF	1	7	8
Method optimization	1	5	9
Orthogonal methods	1	6	8
Reference standard support	3	7	10
Identification and qualification of impurities and degradation products	1	6	8
ICH validation	1	4	10
Technology transfer	1	4	5
Documentation, methods, validation Pkgs, regulatory, etc.	3	6	10
Development and technical reports	6	8	7
Outsourcing	2	7	10
Budget/resources/headcount	2	5	8
Others			

Weighting = 1 Least Critical
= 10 Most Critical

Risk Evaluation Process

Product Development

Risk factors	Weightings (1–10)		
	Early stage	Mid stage	Late stage
Chemical & biological characterization	7	8	10
Formulation characterization	2	8	10
Manufacturability	2	5	10
Excipient compatibility	3	7	10
Formulation development	1	4	10
Clinical plan	10	10	10
Packaging	1	5	10
Scale up & process optimization	1	2	10
Manufacturing site changes	1	4	10
Alternate dosage forms (Technology)	3	6	8
Product definition	1	3	10
Clinical supplies	1	10	10
QC release	1	10	10
Development stability	8	10	10

(continued)

Risk factors	Weightings (1–10)		
	Early stage	Mid stage	Late stage
Commercial process optimization	1	3	10
Technology transfer	1	4	10
Process validation	1	5	10
Registration stability	1	5	10
Commercial stability	1	5	10
Documentation	10	10	10
Outsourcing	10	5	3
Budget	10	10	10
Other			

Weighting   = 1 Least Critical
            = 10 Most Critical

Risk Evaluation

Project Risk Assessment

Risk Evaluation Data Sheet			
Risk Factors	**Weight (1-10) X**	**Rate (1-5) =**	**Risk Value**

Rating   = 1 Most probable to complete   Total Risk Value =
         = 5 least probable to complete

- Activities removed from the current plan are still included in the project risk assessment of the new plan as risk factors and given the highest rating (5)

Risk Evaluation Process

Analytical Plan (Example)

Risk factors	Weightings (1–10)		
	Early stage	Mid stage	Late stage
Implement a project plan		3	
Pre-method development activities			
a. Literature search		3	
b. Obtain chemical and Physiochemical data		8	
c. Collect available samples for method development		10	
d. Determine method criteria based on stage of development		5	
e. Determine chromatography mode		5	
f. Evaluate mobile phase Consideration		6	
Method development			
a. Mobile phase screening and buffer using gradient		10	
b. Column screening		7	
c. Determine whether Isocratic or Gradient		9	
d. Setup solvent selectivity Triangle		10	
e. Run each of the 10 experiments in the selectivity triangle		10	
f. Determine portion of triangle which gives greatest number of peaks and best selectivity		10	
g. Optimize mobile phase		10	
Sample preparation			
a. Determine sample diluent		5	
b. Determine sample solubility		6	
c. Determine analyte stability in diluent		7	
d. Determine nominal analytical concentration (NAC) and injection volume		4	
Method qualification			
a. Linear range		10	
b. Recovery		10	
c. Sensitivity		10	
d. Precision		10	
e. Initial system suitability parameters		10	
f. Development batch analysis		8	
Method validation			
a. Write a draft method document		10	
b. Generate a method validation protocol		10	

(continued)

Risk factors	Weightings (1–10)		
	Early stage	Mid stage	Late stage
c. Approve the method validation protocol		10	
d. Implement validation experiments		10	
Method transfer			
a. Determine type of transfer		3	
b. Prepare intent to transfer document		2	
c. Write transfer protocol		8	
d. Implement transfer analysis		10	
GMP testing		10	
Technical reports			
a. Write an analytical development report		6	
b. Write a method validation report		8	
c. Reports of analysis (C of A)		10	
d. Method transfer report		10	

Risk Evaluation Analytical Plan (Example)

Project Risk Assessment (Total Risk, (R_t), Current Plan)

Risk factors	Weight (1 – 10)	Rate (1 – 5)	Risk value
Risk evaluation data sheet			
Pre method development			
a. Literature search	3	1	3
b. Obtain chemical and Physiochemical data	8	3	24
c. Collect available samples for method development	10	3	30
d. Determine method criteria based on stage of development	5	1	5
e. Determine chromatography mode	5	2	10
f. Evaluate mobile phase consideration	6	2	12
Method development			
a. Mobile phase screening and buffer using gradient	10	2	20
b. Column screening	7	2	14
c. Determine whether Isocratic or Gradient	9	1	9
d. Setup solvent selectivity triangle	10	1	10
e. Run each of the 10 experiments in the selectivity triangle	10	3	30
f. Determine portion of triangle which gives greatest number of peaks and best selectivity	10	2	20
g. Optimize mobile phase	10	2	20

(continued)

Risk evaluation data sheet			
Risk factors	Weight (1 – 10)	Rate (1 – 5)	Risk value
Sample preparation			
a. Determine sample diluent	5	1	5
b. Determine sample solubility	6	2	12
c. Determine analyte stability in diluent	7	3	21
d. Determine nominal analytical concentration (NAC) and injection volume	4	1	4
Method qualification			
a. Linear range	10	2	20
b. Recovery	10	2	20
c. Sensitivity	10	3	30
d. Precision	10	2	20
e. Initial system suitability parameters	10	2	20
f. Development batch analysis	8	2	16
Method validation			
a. Write a draft method document	10	2	20
b. Generate a method validation protocol	10	3	30
c. Approve the method validation protocol	10	2	20
d. Implement validation experiments	10	3	30
Method transfer			
a. Determine type of transfer	3	2	6
b. Prepare intent to transfer document	2	2	4
c. Write transfer protocol	8	2	16
d. Implement transfer analysis	10	3	30
GMP testing	10	3	30
Technical reports			
a. Write an analytical development report	6	3	18
b. Write a method validation report	8	3	24
c. Reports of analysis (C of A)	10	1	10
d. Method transfer report	10	2	20

Total Risk Value (R_t) Current Plan = 633

$$R_t = \frac{\text{Current plan} - \text{Minimum risk}}{\text{Maximum Risk} - \text{Minimum Risk}} \times 100 = \%\text{total Risk}$$

$$\%R_s = 100 - \%R_t = \%\text{probability of success}$$

$$\frac{633-290}{1450-290} \times 100 = \frac{343}{1160} \times 100 = 30\%$$

$$\%\mathbf{R_s} = \mathbf{100\%} - \mathbf{30\%} = \mathbf{70\%}$$

Plan Timing = 53 days Resource (people-days) =110 people-days

Risk Evaluation Analytical Plan (Example)

Project Risk Assessment (Total Risk (R_t), Revised Plan)

Risk evaluation data sheet			
Risk factors	Weight (1 − 10)	Rate (1 − 5)	Risk value
Pre method development			
a. Literature search*	3	5	15
b. Obtain chemical and physiochemical data*	8	5	40
c. Collect available samples for method development	10	4	40
d. Determine method criteria based on stage of development	5	2	10
e. Determine chromatography mode*	5	5	25
f. Evaluate mobile phase Consideration	6	3	18
Method development			
a. Mobile phase screening and buffer using gradient	10	2	20
b. Column screening*	7	5	35
c. Determine whether isocratic or gradient	9	2	18
d. Setup solvent selectivity triangle	10	2	20
e. Run each of the 10 experiments in the selectivity triangle	10	3	30
f. Determine portion of triangle which gives greatest number of peaks and best selectivity	10	2	20
g. Optimize mobile phase	10	2	20
Sample preparation			
a. Determine sample diluent*	5	5	25
b. Determine sample solubility	6	2	12
c. Determine analyte stability in diluent	7	3	21
d. Determine nominal analytical concentration (NAC) and injection volume*	4	5	20
Method qualification*			
a. Linear range	10	5	50
b. Recovery	10	5	50
c. Sensitivity	10	5	50

(continued)

Risk evaluation data sheet			
Risk factors	Weight (1 – 10)	Rate (1 – 5)	Risk value
d. Precision	10	5	50
e. Initial system suitability parameters	10	5	50
f. Development batch analysis	8	3	24
Method validation			
a. Write a draft method document	10	2	20
b. Generate a method validation protocol	10	3	30
c. Approve the method validation protocol	10	2	20
d. Implement validation experiments	10	3	30
Method transfer			
a. Determine type of transfer	3	5	15
b. Prepare intent to transfer document	2	5	10
c. Write transfer protocol	8	2	16
d. Implement transfer analysis	10	3	30
GMP testing	10	4	40
Technical reports			
a. Write an analytical development report	6	3	18
b. Write a method validation report	8	3	24
c. Reports of analysis (C of A)	10	2	20
d. Method transfer report	10	3	30

*Activities removed from the current plan

Total Risk value New Plan = 966

$$R_t = \frac{\text{New plan} - \text{Minimum risk}}{\text{Maximum Risk} - \text{Minimum Risk}} \times 100 = \%\text{total Risk}$$

$$\%R_s = \%100 - \%\text{Total Risk} = \text{probability of success}$$

$$\frac{966 - 290}{1160} \times 100 = \frac{676}{1160} \times 100 = 58\%$$

$$\%R_s = 100 - 58\% = 42\%$$

Plan Timing = 39 days
Resource (people/days) = 87 people-days

Risk Evaluation

The advantages of doing risk analysis:

- Determines risk change from current plan to new plan.
- Presenting risk to other stakeholders (R&D team), which may cause them to change items in their plan, that's driving the risk in your plan.
- Shows the difference between current plan and new plan total risk value, due to requested changes
- Presents probability of success to management.
- Encourages the project team to find alternate solutions for a better plan

Conclusion

The use of the Analytical Project Plan Document is a systematic approach that will allow for a list of all of the activities to be performed, their timing, and their resource commitment required to meet the analytical responsibility. If issues arise that require that the analytical group revise the plan to lower cost and/or reduce project timing, the project plan can be revised to meet the changes required. The advantage of using the project plan approach is that original plan can be directly compared to the revised plan for all stakeholders and management to observe. In addition, a Risk Analysis can be performed for each plan and directly compared to each other. Based on the changes observed in the plans and the risk associated with those changes the responsible individuals can decide whether they are comfortable with the changes and the associated risk. This approach allows for all stakeholders to understand what is involved in the analytical project activities and accept the risk involved in making the changes.

It can be seen from the examples shown above. A current project plan was created which listed all of the activities required to meet the methods intended use. The current plan timing was fairly tight, not leaving room for any errors and re-dos. However, a change was requested to lower the resource cost and shorten the Plans timing. This resulted in having to revise the current plan to meet the requested changes. When the revised plan was developed many activities were removed from the plan to shorten timing and lower resource cost. When the current plan and revised plan were compared to each other, it appeared that the revised plan had all of the essential activities incorporated in it. However, the risks associated with both plans needed to be evaluated. It can be seen by using the project plan document it was easily determined that the timing was shortened from 53 days to 39 days and the resource cost lowered from 110 people-days to 87 people-days. Preforming the Risk Evaluations indicated that the current plan had a 30% risk factor, thereby having a 70% probability of success. While, the revised plan had a 58% risk factor, thus having a 42% probability of success. This approach allows Stakeholders and Management to observe both plans and ask questions, to better understand activities involved in the Analytical Sciences. The Risk Analysis lets them understand the trade off's that occurs between timing, cost, and long-term success.

References

1. Vose D (2008) Risk analysis a quantitative guide, 3rd edn. Wiley, New York
2. Grey S (1995) Practical risk assessment for project management. Wiley series on software practices. Wiley, New York
3. Simon P (1997) Project risk analysis and management. Norwich, APM
4. Catalano T (2013) Essential elements for a GMP analytical chemistry department. Springer, New York

Chapter 6
Project Plan Implementation

Abstract This chapter discusses project plan Implementations which occurs after the project plan is developed and approved. Project Execution Plan and Project Governance Plan flow diagrams must be constructed. In addition, the execution of the plan should be closely monitored. At this stage of the project, activities for project closure implementation should begin.

Governance (Fig. 6.1, Table 6.1)

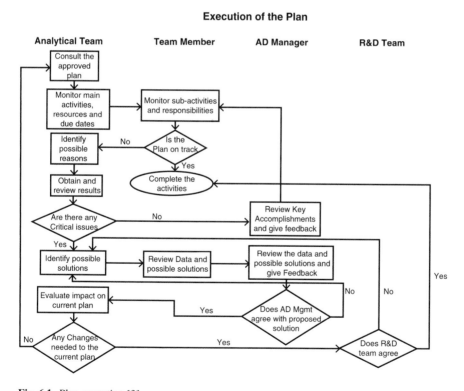

Fig. 6.1 Plan execution [3]

© The Editor(s) (if applicable) and The Author(s), under exclusive license to
Springer Nature Switzerland AG 2020
T. Catalano, *Application of Project Management Principles to the Management of Pharmaceutical R&D Projects*, SpringerBriefs in Pharmaceutical Science & Drug Development, https://doi.org/10.1007/978-3-030-57527-4_6

Table 6.1 Monitor execution of the plan

No.	Key Activity	Duration	Target Start Date	Target Completion Date	Predecessor Activity	Actual Start date	Actual Completion Date	Deviation from Plan	Reason for Deviation	Data Driven Yes/No

The AD management and the analytical team make a presentation at the function project team meeting. The function management responds to the presentation with questions and concerns, AD management captures the concerns and any unanswered questions and instructs the analytical team to evaluate them against the current team project plans. If revisions to the project plans are required, the analytical team will make the necessary revisions and obtain approval from AD management. The analytical team will then schedule another meeting with the function management. If no revisions to the project plans are required the AD management will prepare a cover letter summarizing the meeting conclusions and attaches it to the finalized presentation which is submitted to the function management (Figs. 6.2, 6.3, 6.4, 6.5, 6.6, 6.7 and Table 6.2).

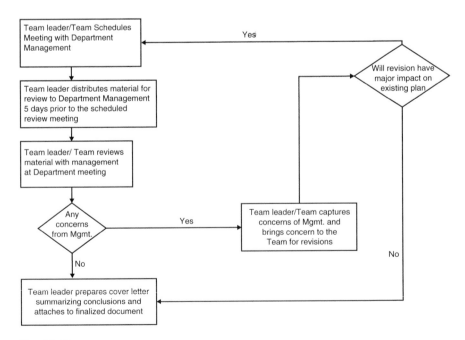

Fig. 6.2 Plan governance

The chart below is comparison of software used for Project Management. The topics shown as heading are the critical activities required to implement Project Management.

Table 6.2 Computer software

Product	Share capabili ties	Applica tion range	Methodol ogies	Budget manage ment	Collabora tion tools	Gantt charts	Mile stone tracking	Portfolio manage ment
Agile	✓	✓	✓	✓	✓	✓	✓	✓
Scrum	✓	✓	✓	✓	✓	✓	✓	✓
Monday.com	✓	✓	✓	✓	✓	✓	✓	✓
Smartsheet	✓	✓	✓	✓	✓	✓	✓	✓
MS project	✓	✓	✓	✓	✓	✓	✓	✓
Trello	✗	✓	✓	✓	✓	✓	✓	✓
Wrike	✓	✓	✓	✓	✓	✓	✓	✓
Waterfront	✗	✓	✓	✓	✓	✓	✓	✓
Function fox	✗	✓	✓	✓	✓	✓	✓	✗

The computer software listed above utilizes the formation of **Information Folders**. Which contain Data, Contact Information, Names, Responsibilities, Comments, Replies, etc. These folders can be moved about, combined, used to communicate with team members, stakeholders, management and used in project meeting presentations. The software also utilizes **Gantt charts** to communicate project implementation activities such as, scheduled start and end times, actual start and end times, responsible person(s), stakeholders, management, resource commitment, precursor activities, comments and replies. The **Project Closure Process** is a strict list of actions that the team, stakeholders and management must follow. All Documentation and Reports must be completed and approved. All contractual obligations have been satisfied, all bills and payments must be finalized. Risk evaluation is done qualitatively and is designated as Low, Medium and High. The Documentation of lessons learned for future reference. Team will be disbanded and members returned to their individual departments.

The paper-based process described herein manages the development and approval of project Strategy and Plans utilizing activity process flow diagrams. **The Project Implementation Process** uses a **Project Execution** activity flow diagram. The plan is monitored by using a spreadsheet listing the metrics to be collected. The use of a **Project Plan Document** allows for a detailed listing of all activities, their start and end times, actual completion date, resource commitment, and precursor links. The governance of project implementation is performed using a **Plan Governance** flow chart. **The Risk Evaluation Process** is a combination of both qualitative determinations and quantitative calculations. Activities are given weighted values based on the stage of development and a rate value based on the probability of completing the task on time (Qualitative). The product of the activity weight value and rate value results in the activities risk value. The sum of all the activities risk values is the projects total risk value which can result in the calculation of projects probability of success (Quantitative). Project Closure is very Similar to that used by the computer software by giving detailed instructions to be followed by team members, stakeholders and management.

Project Closure

Project closure activities should include the recording project documents, archiving all essential project information, ensuring all payments are up to date, releasing resources for transfer to other projects, and completing all the project activities. Every project teaches lessons to the organization whether it's a success or is a failure. So even after a project finishes, the documentation of this project is going to be helpful for completing the coming projects successfully. Below is a guideline that project managers could use to close their projects comprehensively: [1, 3].

1. Ensure that the project has satisfied the goal(s) that were initially set
2. Ensure that the full scope of work has been completed, and obtain formal documentation from the client for the work completed.
3. Review all contracts with the project team and suppliers. Make sure that all parties have satisfied their contractual obligations and that the suppliers have delivered all of the products and/or services required of them. Make sure the supplier provides all deliverables such as, documentation, drawings, warranties, service contracts, and so forth.
4. Ensure all Documentation is completed, approved and achieved
5. Document lessons learned for future reference.
6. Disband the project team and officially return resources to their functional locations.

Documentation is an essential component of Project Management. Without proper documentation, Justification of data generation and data interpretation cannot be obtained. Included in this chapter are templates for several of the important documents that must available for Pharmaceutical R&D. Those described herein are Sample Submission, Technical Reports, Reports of Analysis, and Training Records.

Sample Submission Form		
Company Name		Sample submission tracking sticker placed here
Submission ID Numbers		
Material description	Item Number	Project code
Submitters Name	Date Submitted	
Phone Number	Department Making the Submission	Location
Lot # and other information		
Storage Conditions __Ambient __Refrigerate ___Desiccate ___Freezer ___Protect from Light ___Other_____		
Sample Objective ___R&D Release ___Formal Stability Testing ___Reference Standard ___Tox. Support ___Formulation Support ___Chemical Development Support ___Method Development ___Method Validation ___Formulation Screening ___Other_____		
Testing Requested Appearance Dissolution Assay and Degradation Products Others_____		
Sample Retriever	Date Retrieved	

Fig. 6.3 Sample submission

Development Report
Report Number
Study Name

Version, Date
Page Number

-

Report Number
Report Title
Author

TABLE OF CONTENTS
Sections

Page No.#

1 Purpose & Objective_____
2 Summary_____
3 Discussion_____
4 Conclusion_____
5 References_____

6 LIST of TABLES_____
7 LIST of FIGURES_____

Fig. 6.4 Development report

Company Name Address	Analysis Report Submission ID: Lot #: _____ Material Description:	Database: Approved: Y__ N__ Status: _____		
	Submission Details			
Analysis Goal: (Release, Dev. Support, etc.) Compound Number Project Code Requestors Name:				
	Sample Details			
Sample ID:		User ID:		
	Results Details			
Test Task (e.g. Appearance, Assay, etc.)		Status: _____ Approved: Y___N__		
Component Method Number Analyst Name Notebook Number Notebook Page Instrument Number Compound Name Methodology Units Test -1 Test-2 Test-3 Test-N Test- Mean Test- %RSD	Results	Version	Replicate No#	
			(1, 2, etc.)	
Reviewing Analyst Signature: Printed Name: Approving Supervisor Signature: Printed Name:		Date: Date:		

Fig. 6.5 Analysis report

Stability Report

Table of Contents

1. Summary
    1.1 Purpose of the stability studies
    1.2 Description of material (Drug Substance/ Drug Product)
    1.3 Equipment (Stability Chambers, Stability Data System, etc.)
    1.4 Proposed Shelf Life
    1.5 Guidelines followed
    1.6 Stability Study Protocols
    1.7 Statistical Analysis Utilized

2. Lots used in the stability studies
    2.1 Lots for primary studies (Registration Studies)
    2.2 Lots for Registration like Studies (Phase III studies)
    2.3 Lots for supportive studies (Phase II studies)
    2.4 lots for scale up and Optimization studies (Development lots)

3. Summary of Stability Data
    3.1 Completed data tables for all studies
    3.2 Comparison of all data to specification criteria

4. Evaluation of stability data
    4.1 Identification of data which failed to meet criteria or show negative
        Trends
    4.2 Individual lot Regression Analysis
    4.3 Pooled lot Regression Analysis

5. Photo-stability studies
    5.1 Completed data tables
    5.2 Comparison of all data to specification criteria

6. Stability Conclusion
    6.1 Justification for proposed Shelf Life
    6.2 Statistical Analysis utilized
    6.3 Discussion of any nonconformity to specifications

Approval Page

Author: _____ Date_____

Reviewer: _____ Date_____

Approval: _____ Date_____

Fig. 6.6 Stability report

	Individual Training Record				
Name		Date			
Title					
Curriculam Vitae:					
Professional License or Certificate	Date Issued				
	Compliance Training				
SOP's	2011	2012	2013	2014	2015
PD-10X	√	√	√		
PD-20X	√	√	√		
Guidelines					
GL-10X	√	√	√		
GL-20X	√	√	√		
GMP/GLP	√	√	√		
Courses and Seminars					
Title	Date Taken				
Signature (Staff Member)_____Date_____					
Signature (Compliance Specialist)_____Date_____					

Fig. 6.7 Training record

References

1. Anupindi R et al (1998) Managing business process flows, 2nd edn. Simon and Schuster Custom Publishing, New York
2. Chase RB, Aquilano (1995) Production and operation management. study guide, Amazon. Com. Irwin, New York
3. Portny SE (2017) PMP. Project Management for Dummies, For Dummies; 5 editions

Chapter 7
Financial Project Plan

Abstract Every project team should have a financial management process flow, described in this chapter, for budget planning, budget approval, expense documentation, expense tracking and budget closure. The distinction between unbudgeted and budgeted items must determined and unbudgeted items must be approved by management. The closed budget must be submitted for auditing and approval by management.

Financial Project Plan

A financial project manager oversees projects for companies and corporations that have an impact on the company's revenues. Nearly every firm or organization employs a financial manager to prepare financial reports, create cash management strategies and direct investment activities. In many cases, the project manager plays a key role in developing the long-term financial goals of a company or organization to ensure a profitable future for the firm. The daily tasks of a financial project manager will vary based on the current projects and even the industry of the company [1]. In some industries, he or she will be tasked with the role of controller and oversee the production of financial reports, such as balance sheets, expense reports and income statements. Before a financial project plan can be created the following activities must be considered.

1. Understand the top down approach to setting the project budget. In this circumstance the amount of money allowed to complete the project is predetermined and all project activities cost must be within the budget.
2. Discuss the needs of the project with Team members and Stakeholders, this will allow you to know what the cost of the project will be. This is known as the project scope and defines what are the requirements and the work to be performed.
3. Understand the bottom up approach to setting project budget. Here the cost for each project activity is determined and the total of these costs are combined to determine the project budget.

T. Catalano, *Application of Project Management Principles to the Management of Pharmaceutical R&D Projects*, SpringerBriefs in Pharmaceutical Science & Drug Development, https://doi.org/10.1007/978-3-030-57527-4_7

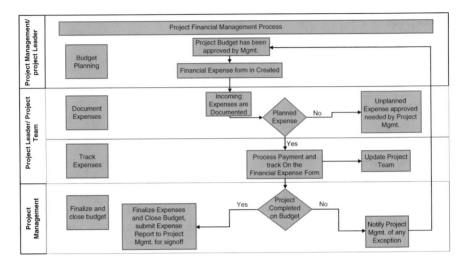

Fig. 7.1 Financial project management Process Flow [2]

Table 7.1 Financial spreadsheet

Expenditure	Cost	Running Totals	Notes

4. Determine the core costs. These are the absolutely essential costs to complete the project. Core costs include items like labor, equipment, materials. The bulk of project costs will come from these components.
5. Non-core expenses. Many projects go over budget because they forget to include non-core expenses. Think of costs outside of labor, equipment, and materials. Consider things like travel, insurance, fuel, and entertainment.
6. Create a table to record your costs. The final aspect of a financial plan is to record all the financial information. It usually consists of four columns labeled as expenditure, costs, running totals, notes (Fig. 7.1 and Table 7.1).

References

1. Clayton M (July 3, 2019) Financial project management, online PM courses; 1 edition
2. Brigham EF, Daves PR (2018) Intermediate financial management, Cengage Learning; 13 editions

Chapter 8
Project Outsourcing

Abstract Small businesses, startup companies and virtual companies must rely heavily on outsourcing to meet time frames and obtain the necessary technical skills to be successful. In this chapter an **Outsourcing Process** is presented which would provide the most efficient and effective approach to outsourcing. Also provided are procedural flow charts and the documentation required for the process.

Project Outsourcing

There are many reasons why companies outsource development activities such as, not having the capabilities in-house, their expertise is limited to a specific area, or may have certain resource capability gaps that require the need for outsourcing. Others may be seeking novel technology or unique capabilities offered by an out-source provider. Tight timelines, competing priorities, or the need to allocate resources for other strategic purposes may also lead to a decision to outsource. In many occasions outsourcing is performed on an individual basis which leads to many concerns such as, higher costs, poor responsiveness, lack of partnership relations etc. [1]. The development of an outsourcing process would provide the most efficient and effective approach to outsourcing.

Flow charts describing the process and the documentation needed are also demonstrated, so that a clear and complete understanding can be obtained (Fig. 8.1).

A critical activity is the selection and qualification of the laboratory to perform the outsourcing testing. Figure 8.2 describes the process for laboratory selection and qualification [2].

© The Editor(s) (if applicable) and The Author(s), under exclusive license to Springer Nature Switzerland AG 2020

T. Catalano, *Application of Project Management Principles to the Management of Pharmaceutical R&D Projects*, SpringerBriefs in Pharmaceutical Science & Drug Development, https://doi.org/10.1007/978-3-030-57527-4_8

Outsource process flow diagram

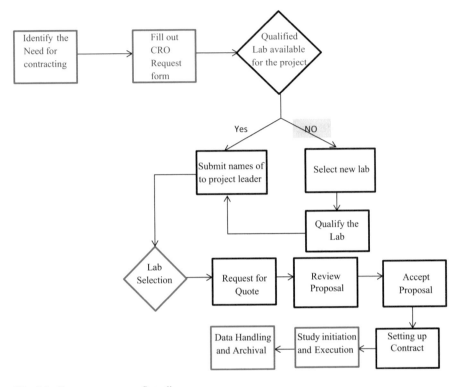

Fig. 8.1 Outsource process flow diagram

Once the project is authorized for contracting and the contracting laboratory is selected (Fig. 8.2), a contacting process is implemented. Figure 8.3 describes the process for contracting.

Within the processes shown above there are several important documents which must be prepared, such as, the Confidentiality Agreement (CDA), Request for Information (RFI), and Request for Quotation. Examples of these documents are shown in below (Figs. 8.4, 8.5 and 8.6):

Laboratory selection and qualification

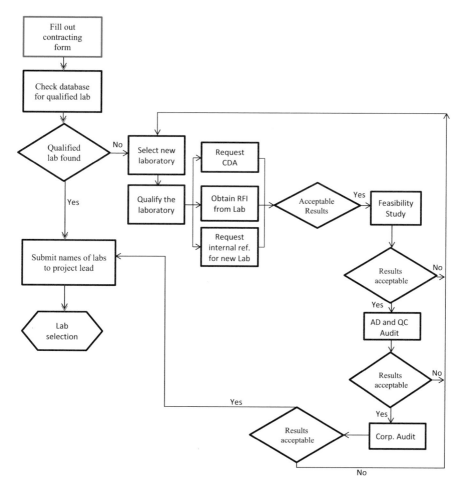

Fig. 8.2 Laboratory selection and qualification

It is essential for the sponsor to establish that the laboratory providing support to the project is reliable and competent. An efficient and effective Audit Program is essential for the qualification of the outsourcing laboratory. There are two types of audits. One, is an on-site audit, which should be performed for all new vendor laboratories. Two, a paper audit which can be performed for approved vendor laboratories [3]. The use of an audit form is a systematic approach to performing a complete and effective audit. A generally used Audit Form is shown below (Figs. 8.7 and 8.8):

Contracting Process

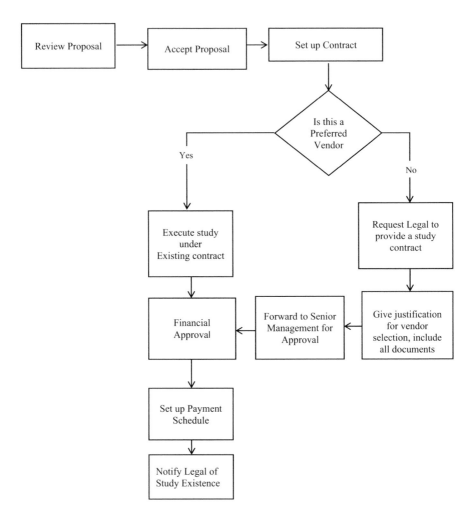

Fig. 8.3 Contracting process

Confidentiality Agreement

Representatives of **Sponsor Inc.**, having a mailing address _____have disclosed or may disclose to Vendor, business information, technical information and/or ideas concerning **Sponsors** ("Proprietary Information").

In consideration of any disclosure of Proprietary Information and for other consideration, the receipt of which is hereby acknowledged, you agree as follows:

1. You will hold in confidence and not possess or use except for purposes of discussions with **Sponsors** representatives (or as otherwise expressly authorized by **Sponsor** in writing) or disclose any Proprietary Information except information you can document (a) is in the public domain through no fault of yours, (b) was known to you prior to disclosure by **Sponsor**, or (c) was properly disclosed to you by another person without restriction. The foregoing does not grant you a license in or to any of the Proprietary Information. Notwithstanding the foregoing, specific aspects of Proprietary Information shall not be deemed to be within the foregoing exceptions when such exceptions apply only to more general knowledge or when the relevant specific aspects are identified using Proprietary Information disclosed under this Agreement.

2. If you are asked by **Sponsor** you will promptly return all Proprietary Information and all copies, extracts, notes and other objects or items in which it may be contained or embodied.

3. You will promptly notify **Sponsor** of any unauthorized release of Proprietary Information.

4. You acknowledge and agree that due to the unique nature of the Proprietary Information, any breach of this Agreement would cause irreparable harm to the **Sponsor** for which damages are not an adequate remedy and that **Sponsor** shall therefore be entitled to equitable relief in addition to all other remedies available at law.

5. The terms of this Agreement will remain in effect with respect to any particular Proprietary Information until you can document that it falls into one of the exceptions stated in Paragraph 1 above.

6. This Agreement is governed by the internal laws of the State of and may be modified or waived only in writing. If any provision is found to be unenforceable, such provision will be limited or deleted to the minimum extent necessary so that the remaining terms remain in full force and effect. You understand that this document does not obligate **Sponsor** to disclose any information or negotiate or enter into any agreement or relationship with you. The prevailing party in any dispute or legal action regarding the subject matter of this Agreement shall be entitled to recover attorneys' fees and costs.

Fig. 8.4 Confidentiality agreement

Acknowledged and agreed on _____, _____.

_SPONSOR   VENDOR

_____   _____
    (Name)   (Name)

By:_____
By:_____

Name (Print) _____   Name (Print)_____

Title: _____
Title:_____

Fig. 8.4 (continued)

Request for Information

1. Company information (General)
   a. Please provide information about your company's, Background, Services, Experience and Qualifications.
   b. Please provide a list of references for companies (minimum 5) for whom you provided analytical and/or stability support services, including contact names, addresses, and a brief description of the services provided.
   c. Do you provide preferred client relationships or strategic partnerships, if yes, please describe salient features of these relationships?
      i. Characterize your current business volume in the current analytical areas
         1. Method Development
         2. QC release testing
         3. Process Support
         4. Registration Stability Studies (FDA, ICH)
         5. Compendial testing
         6. Formulation support
         7. Cleaning validation/verification
         8. Sample storage and management

   d. Does your company possess any specialized expertise or analytical technology, if so please describe them?
   e. Describe your company's management structure and project Management processes, please include an organizational chart and examples of project management work flow process.
   f. Do you perform feasibility studies and co-validations?
   g. Are you willing to engage in multi-year umbrella agreements, so that contracts are not need for every specific project? This would involve having standard terms and conditions in place with a pre-negotiated price structure.
   h. Do you offer cost savings based on volume of business, at what point and under what conditions the cost savings take effect?

2. Facilities
   a. Please describe the site facilities and a list of available equipment
   b. Please provide us with a list of most recent audits (last three years), with a redacted copy of one of your responses, as an example.

Fig. 8.5 Request for information

Request for Information (cont.)

c. Please provide job description for personnel at each level, and the average experience of the personnel.
d. Describe your process for calibration and maintenance of the equipment
e. Describe your document control and retrieval processes
f. Describe your site safety program
g. Describe your site security system
h. Describe facility housekeeping and maintenance program

3. Regulatory Compliance
    a. Describe your Quality Assurance program
        i. Staffing (Number of personnel and qualifications)
        ii. Procedures to ensure compliance with guidelines and regulations for
            1. FDA
            2. Canada
            3. EMA
            4. Japan
        iii. What staff training is available, how is training documented?
        iv. Please furnish a copy of the Table of Contents for your Standard Operating Procedures (SOP)
    b. Please furnish copies of any forms FDA483, warning letters or consent degrees in the last three years, and of your two most recent site Audits

Fig. 8.5 (continued)

Quotation Request		
Requested	Items	Details

	Executive Summary of Project
	  1. Brief Introduction
	  2. Scope of Project	

	Cost for the Work to be Performed
	  1. Feasibility Studies
	  2. Method Development
	  3. Method Validation
	  4. Method Transfer
	  5. Excipient Analysis
	  6. Dosage Form Analysis
	  7. Chemical Analysis
	  8. Raw Material Analysis	

	Equipment Use Cost
	  1. HPLC, GC, Spectroscopy,
	    Microscopy	

	Sample Preparation	
	Set-up Fees	
	Cost of Data Entry	
	Cost of Writing Reports	
	Cost of Re-Analysis	
	Cost of Storage of Samples	
	Cost of Disposal of Samples	

	Stability Study Fees
	  1. Set-up fees
	  2. Testing Fees
	  3. Management Fees	

	Cost for Auditing Technical Data and Compliance Monitoring	

Fig. 8.6 Quotation request

Audit Form

Instructions

1. Complete the Audit Checklist by checking either "Yes" or "No".
2. Identify all non-compliant items ("No" checked items)
3. Write Justifications for non-compliant Items
4. Correct all unjustified non-compliant items
5. Obtain signed report from vendor for all corrections and justifications
6. Attached signed vendor report to GMP audit form
7. Obtain vendor and client Signatures on GMP Audit Form

SOP's and Operating Guidelines

	Yes	No
I. Specifications Development		
II. Documents for Submission to Regulatory Agencies		
III. Planned and Unplanned deviations		
IV. Notebook/Data Handling / Creation and Use		
V. Analysis Request/Sample Handling/Reports of analysis		
VI. Method Validation Packages/Reports		
VII. Method Transfer Process		
VIII. Personnel Training and Certification program		
IX. System Suitability for Chromatographic Methods		
X. Laboratory Investigation of uncharacteristic analytical results		
XI. Equipment Calibration and Maintenance		
XII. GMP Material storage Management		
XIII. Records Retention		
XIV. Retention Sample Policy		

Equipment Calibration and Maintenance

	Yes	No
I. Instrument calibration stickers		
II. Instrument use and maintenance log books		
III. Data system sample Log		
IV. Daily balance check log book		
V. HPLC/GC column use log book		
VI. Review/Approval documentation for vendor performed work		
VII. Instruments for GMP analysis clearly identified		

Chemicals/Reagents/Solutions:

	Yes	No
I. All chemical/reagents/solutions labeled:		
a. Date opened		
b. Date expired		
c. GMP or NON GMP use		
d. Special handling procedures		
II. Separated Acid and Bases storage bins		
III. Clean and Dirty glassware bins clearly identified		

Fig. 8.7 Audit form

Audit Form (cont.)

Sample Receiving and Distribution system

	Yes	No

I. Numbering System for samples
II. Storage of sample awaiting analysis
III. Log book for sample tracking
    a. Submitter name
    b. Date received
    c. Tracking of sample movement
    d. Date dispensed for analysis (analyst name)
    e. Date of sample disposition (disposition)
IV. Analytical Sample retention policy
    a. Sample retention log book
        i. Quantity of sample retained
        ii. Date of sample retention
        iii. Tracking of sample in and out of retention
        iv. Annual review of sample retain process
        v. Standardized report of analysis format

GMP Material Storage Management:

	Yes	No

I. Stability Chambers daily walk through to check on status, During off hours, a daily signing sheet should document time of walk through, name of personnel, and any discrepancies
II. Use and maintenance log-book up to date
III. GMP retains should be stored in an isolated chamber and appropriately labeled
    a. GMP retention log book
        i. Quantity of sample retained
        ii. Date of sample retention
        iii. Tracking of sample in and out of retention
        iv. Annual review of sample retain procedure
IV. All removal of GMP retains need the approval of the Director of QA or designee

Reference Standard Certification Program:

	Yes	No

I. Designation of a Reference Standard Manager
II. Limited access to reference standards (Ref. Std. Mgr. or designee)
III. Reference Standard control numbering system
IV. Reference Standard request form
V. Reference standard tracking system (log book or computer)
VI. Reference Standard Certification and Classification process
VII. Reference Standard Classification usage
VIII. Reference Standard Classification Testing and Criteria
IX. Certification and Recertification Process

Fig. 8.7 (continued)

Audit Form (cont.)

	Yes	No
X. Reference Standard Certificate of Analysis		
XI. Documentation for original material to be certified		
XII. Storage of Reference Standard Material (-20C or lower)		
XIII. Retirement of Reference Standard process		
a. Control of Retains		
XIV. Handling of Reference Standard Material during usage		
a. Aliquot distribution		
b. Return policy		
c. Storage		

GMP Stability Process:

	Yes	No
I. Control of stability chambers (digi-lab, etc.)		
II. Backup system for power failures		
III. Personal notification process for system failures		
IV. Designation of a stability manager		
V. Controlled access to stability area/chambers (Stab. Mgr. or designee)		
VI. Documented access to stability area/chambers		
VII. Tracking of incoming and outgoing samples		
VIII. Preparation of stability samples		
a. GMP facility		
b. Isolated from other laboratory activities		
IX. Protocol generation		
X. Stability results data base		
XI. Stability report generation		
XII. Statistical modeling of data		
a. Determination of retest date		
b. Determination of shelf life		

Personnel Training/Certification Program:

	Yes	No
I. Training records up to date		
II. A documented analyst training/certification program should exist		
a. Analyst trained on all analytical activities regardless of education or experience		
b. All GMP analysis must be done by a trained analyst on that technique		
c. Retraining schedule should be specified		
d. Trainers for each technique should specified		
e. Trainers should attend a train the trainer course		
III. A List of required training for each job description should exist		

Fig. 8.7 (continued)

Audit Form (cont.)

Laboratory Notebooks and Raw Data Control:	Yes	No
I. Designation of notebooks procedure		
II. Table contents		
III. Table of abbreviations		
IV. Formatting of notebook write up		
a. Required information		
b. Legible writing		
c. Use of worksheets (if applicable)		
V. Review and Approval on each page		
VI. Referencing from page to continuation pages		
VII. Notebook review schedule		
a. Data review within time stated in SOP (usually 24 hours)		
b. GMP review within time stated in SOP (usually 3 months)		
VIII. Adjunctive Notebook		
a. Storage of raw data		
b. Individual pages numbered and signed by analyst and reviewer		
c. Cross referenced to Notebook number and page numbers		
IX. Notebooks retention policy		

DATA Systems:	Yes	No
I. Documented Validated Data Systems		
a. Validation meets CFR21 part11 criteria		
b. Scheduled abbreviated re-validation (every 1 or 2 years)		
c. Detailed change control procedure		
d. All changes validated or qualified		
e. Authorized security administrator required for changes to system		
f. All data required approval before released to customers		

Fig. 8.7 (continued)

Audit Report

Company (Vendor) Name	Vendor Company Representative Name /Title	Type of Audit	Audit Document No.	Revision #
Auditing (client) Company name	Client Auditor Name / title		Date	

List all Observations Document #'s, data etc.	**Recommended Observations for Immediate Action**	**initials**
List all Observations Document #'s, data, etc.	**Recommended Observations to be Considered for Action**	**initials**
	Conclusions	

Vendor Signature _____, Date_____

_____

Printed Name

Client Signature_____, Date_____

_____

Printed Name

Fig. 8.8 Audit report

References

1. Hillstrom LC (2013) Outsourcing, reference for business, encyclopedia for business, 2nd edn. New Delhi, Pearson Edition
2. Scully E. Many factors to weigh in decision to outsource, National Underwriters Life and Health Financial Services Edition
3. Catalano T (2013) Essential elements for a GMP analytical chemistry department. Springer, New York

Index

© The Editor(s) (if applicable) and The Author(s), under exclusive license to
Springer Nature Switzerland AG 2020
T. Catalano, *Application of Project Management Principles to the Management of Pharmaceutical R&D Projects*, SpringerBriefs in Pharmaceutical Science & Drug Development, https://doi.org/10.1007/978-3-030-57527-4